Goldbach Conjecture Solution

by

Giorgio Viale

giorgioviale67@gmail.com

Introduction

Goldbach's conjecture is one of the oldest unsolved problems in number theory and all of mathematics.

There are two conjectures, the binomial also called "strong" and the ternary also called "weak".

In 1742 Goldbach wrote a letter to Euler in which he proposed the following conjecture "every odd number greater than five can be written as the sum of three prime numbers" (weak conjecture).

Euler responded by proposing this statement of the conjecture "It states that every even natural number greater than 2 is the sum of two prime numbers" (strong conjecture).

The strong conjecture implies the weak one therefore by proving the first one also proves the second.

Over the centuries various mathematicians and physicists have attempted to find a solution to the two conjectures.

The conjecture has been shown to hold for all integers less than 4×10^{18} but remains unproven despite considerable effort.

For small values of n, the strong Goldbach conjecture (and hence the weak Goldbach conjecture) can be verified directly. For instance, in 1938, Nils Pipping laboriously verified the conjecture up to n = 100000. With the advent of computers, many more values of n have been checked; T. Oliveira e Silva ran a distributed computer search that has verified the conjecture for $n \leq 4 \times 10^{18}$ (and double-checked up to 4×10^{17}) as of 2013. One record from this search is that 3325581707333960528 is the smallest number that cannot be written as a sum of two primes where one is smaller than 9781.

Cully-Hugill and Dudek prove a (partial and conditional) result on the Riemann hypothesis: there exists a sum of two odd primes in the interval $(x, x + 9696 \log^2 x]$ for all $x \geq 2$.

In this text you will find the solution I developed to prove the strong conjecture.

I used the empirical method, starting from Eratosthenes' theorem. This theorem allows us to calculate how many prime numbers there are, given a positive integer n, in the interval included between √n and n. I applied it to pairs of prime numbers.

Let π(2n,√2n) be the function that, given n, yields the number of primes in the interval included between √2n and 2n and let G(2n) be the function that yields the number of pairs of primes p,q in whose sum 2n breaks down.

Since among the elements of these pairs of primes there are primes in the interval included between $\sqrt{2n}$ and $2n$ and since 2,3, ... p, (where p is greatest prime less than or equal to √2n) concur to determinate π(2n,√2n) it is impossible that 2,3, ... p do not concur in the determination of G(2n).

<u>This fact is not the consequence of an axiom or the consequence of a postulate but it is only pure deductive reasoning.</u>

Hence my proof is not based on an axiom or on a postulate but it is purely based on deductive reasoning.

Namely, for every n>1, if there exists a function F(2,3, ... p) such that π(2n,√2n)= F(2,3, ... p) then, for any n>1, there exists a function G(2,3, ... p) such that G(2n)= G(2,3, ... p).

This is a deduction.

Eratosthenes-Legendre Theorem gives the functional form of F and so allowed me to find the functional form of G.

I have to specify that if the functional form of G did not depend from Eratosthenes-Legendre Theorem then Eratosthenes-Legendre Theorem itself would be violated.

In fact functional forms for G not depending from the cited theorem are not allowed.

By the previous deduction.

The Theorem n.2 and its corollaries, which you will find later in the book, highlight how we move from the additive form to the multiplicative one in calculating the number of prime numbers that lie between $\sqrt{2n}$ and $2n$.

Goldbachian Pairs

A goldbachian pair (p,q) is a pair of natural numbers such that p is an odd prime number and q is an odd prime number or 1.

The sum of p and q gives an even number, say 2n.

If n>1 is an integer then one denotes by G(2n) the number of goldbachian pairs for 2n.

Let P be the set of odd prime numbers.

Binary Goldbach's conjecture supposes that for any integer n>1 there exists at least a pair (p,q) in PXPU{1} ("X" is the Cartesian product) such that p+q=2n.

Binary Goldbach's conjecture was a challenging hypothesis, from 2010, for my ambitions.

Begining this text I hoped to mitigate the harshnesses of mathematics unless renounce to theuprightness of their method.

The aim began more and more difficult by the advancing of the works.

Although the result can appear a compromise, there are some tricks to girdle the most hard difficulties.

The proof of the Theorem n.3 is not particularly hard, but a simple study of function can not substitute the reading of this proof.

Theorem n.3 states

$$\left| \prod_{p \leq z}\left(1 - \frac{1}{p}\right) - \frac{e^{-\gamma}}{\ln z} \right| < \frac{1}{(\ln z)^2}$$

Hence setting

$$f(z) = \left| \prod_{p \leq z}\left(1 - \frac{1}{p}\right) - \frac{e^{-\gamma}}{\ln z} \right| - \frac{1}{(\ln z)^2}$$

and observing that $\prod_{p \leq z}\left(1 - \frac{1}{p}\right)$ is a rational number, $\frac{e^{-\gamma}}{\ln z}$ is an irrational number then

the student can apply the necessary notions to study a function to arrive to prove the statement.

Eratosthenes' statement says that the prime numbers included between $\sqrt{n}$ and n are the integers included between $\sqrt{n}$ and n not divisible by 2,3, ... p, where p is greatest prime number less than or equal to $\sqrt{n}$. (See also Eratosthenes-Legendre Theorem in the paragraph "Preliminary results").

Let $S(\sqrt{n})$ be the set constituted by 2,3, ... p, where p is greatest prime number less than or equal to $\sqrt{n}$.

Let j be an integer included within 0 and n.

Then the pair (n-j,n+j) such that j is incongruent to n modulo p and such that j is incongruent to -n modulo p where p is the generic element of $S(\sqrt{n})$ is a goldbachian pair for 2n.

In formulae we have

$|A| = |\{j \ s.t. \ 0 \leq j \leq n \ s.t. \ \forall p \in S(\sqrt{n}) \ j \neq n \bmod p \ \text{and} \ j \neq -n \bmod p \}| \leq G(2n)$

To prove this statement observe that

- By Eratosthenes' statement both n-j and n+j are prime numbers;
- (n-j)+(n+j)=2n.

In the following I shall denote the set A by "S".

Given n=10 the reader is invited to find how many goldbachian pairs there are for 2n and how many goldbachian pairs there are for 2n in S.

Euler's Totient Function $\varphi(n)$ gives the number of integers k included within 1 and n such that

Gcd(k,n)=1. Let P be the set of prime numbers. We have

$$\varphi(n) = n \prod_{\substack{p|n \\ p \in P}} \left(1 - \frac{1}{p}\right) \quad (A)$$

Two different proofs of (A) are contained in the books of David Burton and Tom Apostol.

(See the Bibliography)

Given the prime factorization of the positive integer n, that is

$$n = \prod_{\substack{p|n \\ p \in P}} p^{\alpha(p)}$$

Mobius μ Function is defined as

- $\mu(1) = 1$;
- $\mu(n) = (-1)^{|\{p \in P: p|n\}|}$ if $\forall p\ \alpha(p) = 1$
- $\mu(n) = 0$ otherwise

The formula

$\varphi(n) = \sum_{d|n} \mu(d) \frac{n}{d}$ (B)

used in the preliminary observation to the proof of Theorem n.4 is proven in the books of David Burton and Tom Apostol. (See the Bibliography)

The formula

$\varphi(n) = \sum_{d|n} \mu(d) \frac{n}{d}$ (B)

used in the preliminary observation to the proof of Theorem n.4 is proven in the books of David Burton and Tom Apostol.

The reader can also find a proof of (A) and of (B) in the Appendix of this text.

This text contains:

1) A brief on the notation I usually use.

2) Some preliminary results:

    Theorem n.1 (Eratosthene-Legendre);

    Theorem n.2;

    Theorem n.3;

3) Goldbach's conjecture solution.

4) A conclusion

5) An Appendix

6) A bibliography.

I recall that Mertens' Theorem states:
$$\prod_{p \leq z}\left(1-\frac{1}{p}\right) = \frac{e^{-\gamma}}{\log z} \cdot \left(1 + O\left(\frac{1}{\log z}\right)\right)$$
In a similar way Theorem n.3 states
$$\left|\prod_{p \leq z}\left(1-\frac{1}{p}\right) - \frac{e^{-\gamma}}{\ln z}\right| < \frac{1}{(\ln z)^2}$$
Since Theorem n.3 implies Mertens' Theorem, but the converse is not true, I have decided to prove the Theorem n.3.

Notation

By P I shall denote the set of prime number.

By p, q and r I shall denote the generic elements of P.

Let x be a real number. By $[x]$

I shall denote the integer part of x.

By $\overline{n}$ I shall denote the discrete set,

that is

$\overline{n} = \{0,1,2,...n-2, n-1, n\}$.

Therefore $\overline{[\sqrt{5}]} = \{0,1,2\}$ and $\overline{[\sqrt{5}]} \cap P = \{2\}$.

Let n, k be two integers.

If k divides n then I shall write k|n;

otherwise I shall write k∤n.

If n, k are not both zero, by the symbol (n, k)

I shall denote the greatest common divisor of k and n.

Let $n = p_1^{e_1} \cdot p_2^{e_2} \cdot .... \cdot p_r^{e_r}$ be the prime factorization of n.

Then Mobius μ function definition is:

$\mu(1) = 1$, $(-1)^r = \mu(n)$ if $e_1 = e_2 = ... e_r = 1$, $0 = \mu(n)$ otherwise.

As usually I set

$G(2 \cdot n) = |\{(p,q) \in PXP \bigcup \{1\} : p + q = 2 \cdot n\}|$

Recall that

$$\gamma = \lim_{x \to \infty} \left(\sum_{n \leq x} \frac{1}{n} - \log x \right) < \infty$$

In fact

$$\sum_{n=2}^{x} \frac{1}{n} - \log x = \sum_{n=2}^{x} \frac{1}{n} - \int_{1}^{x} \frac{dt}{t} = \sum_{n=2}^{x} \left(\frac{1}{n} - \int_{n-1}^{n} \frac{dt}{t} \right) = \sum_{n=2}^{x} \left(\frac{1}{n} - \log\left(1 - \frac{1}{n}\right) \right)$$

and

$$\frac{1}{n} - \log\left(1 - \frac{1}{n}\right) = O\left(\frac{1}{n^2}\right)$$

Since

$$\sum_{n=1}^{\infty} \frac{1}{n^2} = \frac{\pi^2}{6}, \quad \sum_{n=2}^{x} \left(\frac{1}{n} - \log\left(1 - \frac{1}{n}\right) \right) \text{ converges to a limit as } x \to \infty.$$

Set now $0 \leq \varepsilon < 1, n \geq 2, M = n + \lim_{n \to \infty} \frac{\varepsilon}{n}$.

Since

$$\frac{1}{2} + \frac{1}{3} + \frac{1}{4} + \ldots + \frac{1}{M} \leq \ln M \leq 1 + \frac{1}{2} + \frac{1}{3} + \frac{1}{4} + \ldots + \frac{1}{M-1}$$

we have

$$\frac{1}{n} \leq 1 + \frac{1}{2} + \frac{1}{3} + \frac{1}{4} + \ldots + \frac{1}{n} - \ln M \leq 1$$

Hence $\gamma \in [0,1]$.

By trial $\gamma = 0,57\ldots$

Preliminary results

Theorem n.1. Let P_z be the product of the primes $p \leq z$, $\pi(x,z)$ the number of $n \leq x$ that are not divisible by any prime $p \leq z$. Then

$$\pi(x,z) = \sum_{d|P_z} \mu(d) \left[\frac{x}{d}\right]$$

Proof of Theorem n.1.

$$\pi(x,z) = \sum_{n \leq x} \sum_{d|(n,P_z)} \mu(d) = \sum_{d|P_z} \mu(d) \sum_{\substack{n \leq x \\ d|n}} 1 = \sum_{d|P_z} \mu(d) \left[\frac{x}{d}\right]$$

Definition n.1. Let n,k two integers not both zero.

An arithmetical function f is a multiplicative function if $(n,k) = 1$ then $f(n \cdot k) = f(n) \cdot f(k)$.

Theorem n.2. Let f be multiplicative. Suppose that

$$n = \prod_{p^\alpha \| n} p^\alpha$$

is the unique factorization of n into powers of distinct primes. Then

$$\sum_{d|n} f(d) = \prod_{p^\alpha \| n} \left(1 + f(p) + f(p^2) + \ldots + f(p^\alpha)\right)$$

Proof of Theorem n.2.

By multiplicativity of f

$$\sum_{k|n} f(k) = \sum_{d|n} f(d) \cdot \sum_{\substack{e|\frac{n}{d} \\ (d,e)=1}} f(e)$$

Write then

$$\prod_{p^\alpha \| n} p^\alpha = \prod_{i=1}^{r} p_i^{\alpha_i} = \prod_{i=1}^{r} P_i$$

and define

$$S_i = \frac{n}{P_i}$$

to obtain

$$\sum_{d|n} f(d) = \sum_{d|P_i} f(d) \cdot \sum_{e|\frac{n}{P_i}} f(e)$$

$$= \sum_{d_1|P_1} f(d_1) \cdot \sum_{d_2|\frac{n}{S_2}} f(d_2) \cdot \sum_{d_3|\frac{n}{S_3}} f(d_3) \cdot \ldots \cdot \sum_{1|n} f(1)$$

Observe now that

$$\sum_{d_1|P_1} f(d_1) \cdot \sum_{d_2|\frac{n}{S_2}} f(d_2) \cdot \sum_{d_3|\frac{n}{S_3}} f(d_3) \cdot \ldots \cdot \sum_{1|n} f(1)$$

$$= \prod_{p^\alpha \| n} \left(1 + f(p) + f(p^2) + \ldots + f(p^\alpha)\right)$$

Hence

$$\sum_{d|n} f(d) = \prod_{p^\alpha \| n} \left(1 + f(p) + f(p^2) + \ldots + f(p^\alpha)\right)$$

First Corollary to Theorem n.2.

$$\sum_{d|n} \frac{\mu(d)}{d} = \prod_{p|n}\left(1 - \frac{1}{p}\right)$$

Proof of the first Corollary to Theorem n.2.
By definition of μ.

Second Corollary to Theorem n.2.

$$n \geq 4 \Rightarrow \forall n \sum_{d|P_{\sqrt{n}}} \mu(d) \cdot \left[\frac{n}{d}\right] = \left[\sum_{d|P_{\sqrt{n}}} \mu(d) \cdot \frac{n}{d}\right] = \left[n \cdot \prod_{p|P_{\sqrt{n}}}\left(1 - \frac{1}{p}\right)\right]$$

Proof of the second Corollary to Theorem n.2.
We can rewrite the statement as

$$n \geq 4 \Rightarrow \forall n \left[\sum_{d|P_{\sqrt{n}}} \mu(d) \cdot \frac{n}{d}\right] - \sum_{d|P_{\sqrt{n}}} \mu(d) \cdot \left[\frac{n}{d}\right] = 0$$

Now write
$$\left\{\frac{n}{d}\right\} = \frac{n}{d} - \left[\frac{n}{d}\right]$$
so that we can prove

$$n \geq 4 \Rightarrow \forall n \sum_{d|P_{\sqrt{n}}} \mu(d) \cdot \frac{n}{d} - \left\{\sum_{d|P_{\sqrt{n}}} \mu(d) \cdot \frac{n}{d}\right\} - \left(\sum_{d|P_{\sqrt{n}}} \mu(d) \cdot \left(\frac{n}{d} - \left\{\frac{n}{d}\right\}\right)\right) = 0$$

or

$$n \geq 4 \Rightarrow \forall n \sum_{d|P_{\sqrt{n}}} \mu(d) \cdot \left\{\frac{n}{d}\right\} - \left\{\sum_{d|P_{\sqrt{n}}} \mu(d) \cdot \frac{n}{d}\right\} = 0$$

Recalling that $0 \leq \left\{\sum_{d|P_{\sqrt{n}}} \mu(d) \cdot \frac{n}{d}\right\} < 1$ we can prove that

$$n \geq 4 \Rightarrow \forall n \sum_{d|P_{\sqrt{n}}} \mu(d) \cdot \left\{\frac{n}{d}\right\} - \left\{\sum_{d|P_{\sqrt{n}}} \mu(d) \cdot \frac{n}{d}\right\} = 0$$

by induction on the number of prime divisors of $P_{\sqrt{n}}$.

When $P_{\sqrt{n}}$ has only one prime divisor p (namely p=2) there exists a $\delta \geq 1$ such that
we have

$$\sum_{d|P_{\sqrt{n}}} \mu(d) \cdot \left\{\frac{n}{d}\right\} - \left\{\sum_{d|P_{\sqrt{n}}} \mu(d) \cdot \frac{n}{d}\right\} = -\left\{\frac{n+\delta}{p}\right\} - \left\{1 - \frac{n+\delta}{p}\right\}$$

$$= \left\{\frac{n+\delta}{p}\right\} - \left\{\frac{n+\delta}{p}\right\} = 0$$

Let p be the generic prime divisor of $P_{\sqrt{n}}$ such that p does not divide n.

Assume now inductively that whenever $P_{\sqrt{n}}$ has $k > 1$ prime divisors p we have

$$n \geq 4 \Rightarrow \forall n \; 0 \leq \sum_{d|P_{\sqrt{n}}} \mu(d) \cdot \left\{\frac{n}{d}\right\} - \left\{\sum_{d|P_{\sqrt{n}}} \mu(d) \cdot \frac{n}{d}\right\} \leq 1$$

and try to prove that, whenever $P_{\sqrt{n+\delta}}$ has k+1 prime divisors p say $\{p_1, \dots p_k, p_{k+1}\}$ and $p_{k+1} \nmid n+\delta$, we have

$$n \geq 4 \Rightarrow \forall n \; \sum_{e|P_{\sqrt{n+\delta}}} \mu(e) \cdot \left\{\frac{n+\delta}{e}\right\} - \left\{\sum_{e|P_{\sqrt{n+\delta}}} \mu(e) \cdot \frac{n+\delta}{e}\right\} = 0, \; \delta \geq (p_{k+1} - p_k)^2$$

Recall now that inductive hypothesis implies that

$$-\prod_{p|P_{\sqrt{n}}}\left\{\frac{n}{p}\right\} = \prod_{p|P_{\sqrt{n}}} \mu(p) \cdot \frac{n \bmod p}{p} = \sum_{d|P_{\sqrt{n}}} \mu(d) \cdot \frac{n \bmod d}{d} = \sum_{d|P_{\sqrt{n}}} \mu(d) \cdot \left\{\frac{n}{d}\right\} = \left\{\sum_{d|P_{\sqrt{n}}} \mu(d) \cdot \frac{n}{d}\right\}$$

Since

$$n \geq 4 \Rightarrow \forall n \; \sum_{e|P_{\sqrt{n+\delta}}} \mu(e) \cdot \frac{n+\delta}{e}$$

$$= \sum_{d|P_{\sqrt{n}}} \mu(d) \cdot \frac{n+\delta}{d} + \sum_{d|P_{\sqrt{n}}} \mu(p_{k+1} \cdot d) \cdot \frac{n+\delta}{p_{k+1} \cdot d} =$$

$$\sum_{d|P_{\sqrt{n}}} \mu(d) \cdot \frac{n+\delta}{d} \cdot \left(1 + \frac{\mu(p_{k+1})}{p_{k+1}}\right)$$

$$= \sum_{d|P_{\sqrt{n}}} \mu(d) \cdot \frac{n}{d} \cdot \left(1 + \frac{\delta}{n}\right) \cdot \left(1 + \frac{\mu(p_{k+1})}{p_{k+1}}\right) =$$

$$\sum_{d|P_{\sqrt{n}}} \mu(d) \cdot \frac{n}{d} \cdot \left(1 - \frac{1}{p_{k+1}} + \frac{\delta}{n} - \frac{\delta}{p_{k+1} \cdot n}\right)$$

we have

$$n \geq 4 \Rightarrow \forall n \sum_{e | P_{\sqrt{n+\delta}}} \mu(e) \cdot \left\{ \frac{n+\delta}{e} \right\}$$

$$= \sum_{d | P_{\sqrt{n}}} \mu(d) \cdot \left\{ \frac{n+\delta}{d} \right\} + \sum_{d | P_{\sqrt{n}}} \mu(p_{k+1} \cdot d) \cdot \left\{ \frac{n+\delta}{p_{k+1} \cdot d} \right\} =$$

$$\sum_{d | P_{\sqrt{n}}} \mu(d) \cdot \left\{ \frac{n}{d} \right\} \cdot \left(1 - \frac{1}{p_{k+1}} + \frac{\delta}{n} - \frac{\delta}{p_{k+1} \cdot n} \right)$$

$$= -\prod_{p | P_{\sqrt{n}}} \left\{ \frac{n}{p} \right\} \cdot \left(1 - \frac{1}{p_{k+1}} + \frac{\delta}{n} - \frac{\delta}{p_{k+1} \cdot n} \right)$$

Then observe that

$$n \geq 4 \Rightarrow \forall n \left\{ \sum_{e | P_{\sqrt{n+\delta}}} \mu(e) \cdot \frac{n+\delta}{e} \right\}$$

$$= \left\{ \sum_{d | P_{\sqrt{n}}} \mu(d) \cdot \frac{n+\delta}{d} + \sum_{d | P_{\sqrt{n}}} \mu(p_{k+1} \cdot d) \cdot \frac{n+\delta}{p_{k+1} \cdot d} \right\} =$$

$$\left\{ \sum_{d | P_{\sqrt{n}}} \mu(d) \cdot \frac{n+\delta}{d} - \sum_{d | P_{\sqrt{n}}} \mu(d) \cdot \frac{n+\delta}{p_{k+1} \cdot d} \right\}$$

$$= \left\{ \sum_{d | P_{\sqrt{n}}} \mu(d) \cdot \frac{n+\delta}{d} \cdot \left(1 - \frac{1}{p_{k+1}} \right) \right\} =$$

$$= \left\{ \sum_{d | P_{\sqrt{n}}} \mu(d) \cdot \frac{n}{d} \cdot \left(1 - \frac{1}{p_{k+1}} + \frac{\delta}{n} - \frac{\delta}{p_{k+1} \cdot n} \right) \right\} =$$

$$\left(1 - \frac{1}{p_{k+1}} + \frac{\delta}{n} - \frac{\delta}{p_{k+1} \cdot n} \right) \cdot \left\{ \sum_{d | P_{\sqrt{n}}} \mu(d) \cdot \frac{n}{d} \right\}$$

so that

$$n \geq 4 \Rightarrow \forall n \sum_{e|P_{\sqrt{n+\delta}}} \mu(e) \cdot \left\{\frac{n+\delta}{e}\right\} - \left\{\sum_{e|P_{\sqrt{n+\delta}}} \mu(e) \cdot \frac{n+\delta}{e}\right\} =$$

$$\left(1 - \frac{1}{p_{k+1}} + \frac{\delta}{n} - \frac{\delta}{p_{k+1} \cdot n}\right) \cdot \sum_{d|P_{\sqrt{n}}} \mu(d) \cdot \left\{\frac{n}{d}\right\} - \left(1 - \frac{1}{p_{k+1}} + \frac{\delta}{n} - \frac{\delta}{p_{k+1} \cdot n}\right) \cdot \left\{\sum_{d|P_{\sqrt{n}}} \mu(d) \cdot \frac{n}{d}\right\} =$$

$$\left(1 - \frac{1}{p_{k+1}} + \frac{\delta}{n} - \frac{\delta}{p_{k+1} \cdot n}\right) \cdot \left(\sum_{d|P_{\sqrt{n}}} \mu(d) \cdot \left\{\frac{n}{d}\right\} - \left\{\sum_{d|P_{\sqrt{n}}} \mu(d) \cdot \frac{n}{d}\right\}\right) =$$

$$\left(1 - \frac{1}{p_{k+1}} + \frac{\delta}{n} - \frac{\delta}{p_{k+1} \cdot n}\right) \cdot \left(-\prod_{p|P_{\sqrt{n}}} \left\{\frac{n}{p}\right\} - \left\{\sum_{d|P_{\sqrt{n}}} \mu(d) \cdot \frac{n}{d}\right\}\right) =$$

$$\left(1 - \frac{1}{p_{k+1}} + \frac{\delta}{n} - \frac{\delta}{p_{k+1} \cdot n}\right) \cdot \left(\prod_{p|P_{\sqrt{n}}} \left\{\frac{n}{p}\right\} - \prod_{p|P_{\sqrt{n}}} \left\{\frac{n}{p}\right\}\right) \text{ by inductive hypothesis}$$

$$= \left(1 - \frac{1}{p_{k+1}} + \frac{\delta}{n} - \frac{\delta}{p_{k+1} \cdot n}\right) \cdot 0 = 0$$

Hence

$$n \geq 4 \Rightarrow \forall n \sum_{e|P_{\sqrt{n+\delta}}} \mu(e) \cdot \left\{\frac{n+\delta}{e}\right\} - \left\{\sum_{e|P_{\sqrt{n+\delta}}} \mu(e) \cdot \frac{n+\delta}{e}\right\} = 0$$

and our induction is complete.

Third Corollary to Theorem n.2.

$$n \geq 4 \wedge e \in P \wedge e \nmid n \wedge e \mid P_{\sqrt{n}} \Rightarrow \forall n \exists e \ 0 < \sum_{e \mid P_{\sqrt{n}}} \mu(e) \cdot \frac{n}{e} - \sum_{e \mid P_{\sqrt{n}}} \mu(e) \cdot \left[\frac{n}{e}\right] < 1$$

Proof of the third Corollary to Theorem n.2.

I give a proof by contradiction. Suppose that

$$n \geq 4 \wedge e \in P \wedge e \nmid n \wedge e \mid P_{\sqrt{n}} \Rightarrow \forall n \exists e \ \sum_{e \mid P_{\sqrt{n}}} \mu(e) \cdot \frac{n}{e} - \sum_{e \mid P_{\sqrt{n}}} \mu(e) \cdot \left[\frac{n}{e}\right]$$

$$= \sum_{e \mid P_{\sqrt{n}}} \mu(e) \cdot \left\{\frac{n}{e}\right\} > 1$$

By the proof of the second Corollary of Theorem n.2 we have

$$n \geq 4 \wedge e \in P \wedge e \nmid n \wedge e \mid P_{\sqrt{n}} \Rightarrow \forall n \exists e \ 1 < \sum_{e \mid P_{\sqrt{n}}} \mu(e) \cdot \left\{\frac{n}{e}\right\}$$

$$= \left\{\sum_{e \mid P_{\sqrt{n}}} \mu(e) \cdot \frac{n}{e}\right\} < 1$$

that implies $n \geq 4 \wedge e \in P \wedge e \nmid n \wedge e \mid P_{\sqrt{n}} \Rightarrow \forall n \exists e \ 1 < 1$

The contradiction we have reached concludes the proof.

Theorem n. 3. Let be $p \in P$ and let $z \geq 2$ be a real number. Then

$$\left|\frac{e^{-\gamma}}{\ln z} - \prod_{p \leq z}(1-\frac{1}{p})\right| < \frac{1}{(\ln z)^2}$$

Proof of the Theorem n. 3.
Since

$$\lim_{z \to \infty} \frac{e^{-\gamma}}{\ln z} - \prod_{p \leq z}(1-\frac{1}{p}) = 0 = \lim_{z \to \infty} \frac{1}{(\ln z)^2}$$

and since $\lim_{z \to \infty}(\ln z)^2 \cdot e^{-z} = 0$ in $[2, \infty[$ there exists a λ such that

$$\lim_{z \to \infty}(\ln z)^2 \cdot e^{-\lambda \cdot z} \cdot \left|1 - \frac{\prod_{p \leq z}(1-\frac{1}{p})}{e^{\lambda \cdot z}}\right| = 0, \lim_{z \to \infty}(\ln z)^2 \cdot e^{-\lambda \cdot z} \cdot \left|1 - \frac{1}{\ln z \cdot e^{\lambda \cdot z + \gamma}}\right| = 0$$

so that

$$\left|\prod_{p \leq z}(1-\frac{1}{p}) - e^{-\lambda \cdot z}\right| = o\left(\frac{1}{(\ln z)^2}\right) \text{ and so that } \left|e^{-\lambda \cdot z} - \frac{e^{-\gamma}}{\ln z}\right| = o\left(\frac{1}{(\ln z)^2}\right).$$

Therefore since

$$|\prod_{p\leq z}(1-\frac{1}{p})-e^{-\lambda\cdot z}|=o\left(\frac{1}{(\ln z)^2}\right) \text{ and } |e^{-\lambda\cdot z}-\frac{e^{-\gamma}}{\ln z}|=o\left(\frac{1}{(\ln z)^2}\right) \text{ we have}$$

$$|\prod_{p\leq z}(1-\frac{1}{p})-e^{-\lambda\cdot z}|+|e^{-\lambda\cdot z}-\frac{e^{-\gamma}}{\ln z}|=|\prod_{p\leq z}(1-\frac{1}{p})-\frac{e^{-\gamma}}{\ln z}|=o\left(\frac{1}{(\ln z)^2}\right)$$

and

$$\lim_{z\to\infty}\frac{|\prod_{p\leq z}(1-\frac{1}{p})-\frac{e^{-\gamma}}{\ln z}|}{\frac{1}{(\ln z)^2}}=\lim_{z\to\infty}(\ln z)^2\cdot|\prod_{p\leq z}(1-\frac{1}{p})-\frac{e^{-\gamma}}{\ln z}|=0$$

Now observe that if there existed a $z \in \mathbb{N}$, $z \geq 2$,

such that $|\prod_{p\leq z}(1-\frac{1}{p})-\frac{e^{-\gamma}}{\ln z}|\geq \frac{1}{(\ln z)^2}$ the function

$$\delta=|\prod_{p\leq z}(1-\frac{1}{p})-\frac{e^{-\gamma}}{\ln z}|-\frac{1}{(\ln z)^2}=|\prod_{p\leq z}(1-\frac{1}{p})-\frac{e^{-\gamma}}{\ln z}-\frac{1}{(\ln z)^2}|$$

would have at least a change of sign,

so that in this case there should exist at least a $z \in \mathbb{N}$, $z \geq 2$, such that $\delta=0$.

This is impossible because

$\forall z$ the pairs $\left(z, \prod_{p\leq z}(1-\frac{1}{p})\right)$ are pairs of rational numbers

and the pairs $\left(z, \frac{e^{-\gamma}}{\ln z}+(\ln z)^{-2}\right)$ are pairs of a rational number and an irrational number.

This observation concludes the proof of Theorem n.3.

Remark n.1.

The proof can also be set up observing that there exists an initial value (i.e. $z = 2$) such that the Theorem holds. Our aim is now to prove that does not exist a $z \in]2,\infty[$ such that

$$\frac{1}{(\ln z)^2} \cdot \left(1 - e^{-\gamma} \cdot \ln z + \prod_{p \leq z}(1 - \frac{1}{p}) \cdot (\ln z)^2\right) = 0$$

and that does not exist a $z \in]2,\infty[$ such that

$$\frac{1}{(\ln z)^2} \cdot \left(1 + e^{-\gamma} \cdot \ln z - \prod_{p \leq z}(1 - \frac{1}{p}) \cdot (\ln z)^2\right) = 0$$

We can complete the proof observing that

$$\frac{1}{(\ln z)^2} \cdot \left(1 - e^{-\gamma} \cdot \ln z + \prod_{p \leq z}(1 - \frac{1}{p}) \cdot (\ln z)^2\right) = 0 \Rightarrow e^{-\gamma} \cdot \ln z - \prod_{p \leq z}(1 - \frac{1}{p}) \cdot (\ln z)^2 = 1$$

and

$$\frac{1}{(\ln z)^2} \cdot \left(1 + e^{-\gamma} \cdot \ln z - \prod_{p \leq z}(1 - \frac{1}{p}) \cdot (\ln z)^2\right) = 0 \Rightarrow \prod_{p \leq z}(1 - \frac{1}{p}) \cdot (\ln z)^2 - e^{-\gamma} \cdot \ln z = 1$$

Since $\left(z, e^{-\gamma} \cdot \ln z - \prod_{p \leq z}(1 - \frac{1}{p}) \cdot (\ln z)^2\right) \in QXR \setminus Q$ and $(z,1) \in QXQ$, the equation

$e^{-\gamma} \cdot \ln z - \prod_{p \leq z}(1 - \frac{1}{p}) \cdot (\ln z)^2 = 1$ is impossible in $z \in]2,\infty[$.

An analogous reasoning is valid for the second equation.

Problems

Let P_z be the product of the primes $p \leq z$, $\pi(x,z)$ the number of $n \leq x$ that are not divisible by any prime $p \leq z$ and $\pi(z)$ the number of primes $q \leq z$.

Prove that
$$n \geq 4 \Rightarrow \forall n \; \pi(\sqrt{n}) \leq \pi(n,\sqrt{n}) = \sum_{d | P_{\sqrt{n}}} \mu(d) \left[\frac{n}{d}\right]$$

Solution.

We have
$$n \geq 4 \Rightarrow \forall n \; \frac{\pi(\sqrt{n})}{n} \leq \frac{1}{n} \cdot \frac{\sqrt{n}}{2} = \frac{1}{2} \cdot \frac{1}{\sqrt{n}} \leq \frac{1}{2} \cdot \prod_{p | P_{\sqrt{n}}} \left(1 - \frac{1}{p}\right) = \frac{1}{2} \cdot \sum_{d | P_{\sqrt{n}}} \frac{\mu(d)}{d}$$

Hence multiplying LHS and RHS by n we have
$$n \geq 4 \Rightarrow \forall n \; \pi(\sqrt{n}) \leq \frac{n}{2} \cdot \sum_{d | P_{\sqrt{n}}} \frac{\mu(d)}{d} \leq \sum_{d | P_{\sqrt{n}}} \mu(d) \left[\frac{n}{d}\right] = \pi(n,\sqrt{n})$$

since $n > 0 \Rightarrow \forall n \; \sum_{d | P_{\sqrt{n}}} \mu(d) \left[\frac{n}{d}\right] \leq \sum_{d | P_{\sqrt{n}}} \mu(d) \left(\left[\frac{n}{d}\right] + \left\{\frac{n}{d}\right\}\right)$

and $\sum_{d | P_{\sqrt{n}}} \mu(d) \left\{\frac{n}{d}\right\} \in [0,1]$ by the proof of the second Corollary to Theorem n.1.

Prove that

$$n > 1 \Rightarrow \forall n \sum_{d | P_{\sqrt{n}}} \frac{\mu(d)}{d} = \prod_{\substack{p \leq \sqrt{n} \\ p | 2 \cdot n}} \left(1 - \frac{1}{p}\right) \cdot \prod_{\substack{p \leq \sqrt{n} \\ p \nmid 2 \cdot n}} \left(1 - \frac{1}{p}\right) \geq \frac{1}{\sqrt{2 \cdot n}}$$

Solution.

Let $n > 1$. Writing

$$n = 2^e \cdot \prod_{\substack{p | n \\ p \neq 2}} p^{\alpha(p)}$$

so that

$$\varphi(n) = \frac{1}{2} \cdot 2^e \cdot \prod_{\substack{p | n \\ p \neq 2}} (p-1) \cdot \prod_{\substack{p | n \\ p \neq 2}} p^{\alpha(p)-1}$$

and exploiting the inequalities $p - 1 > \sqrt{p}, \alpha(p) - \frac{1}{2} \geq \frac{\alpha(p)}{2}$ we have

$$\varphi(n) = \frac{1}{2} \cdot 2^e \cdot \prod_{\substack{p | n \\ p \neq 2}} (p-1) \cdot \prod_{\substack{p | n \\ p \neq 2}} p^{\alpha(p)-1} \geq \frac{1}{2} \cdot 2^{\frac{e}{2}} \cdot p^{\frac{1}{2}} \cdot p^{\frac{\alpha(p)}{2} - \frac{1}{2}} = \frac{1}{2} \cdot 2^{\frac{e}{2}} \cdot p^{\frac{\alpha(p)}{2}} = \frac{1}{2} \cdot \sqrt{2 \cdot n}$$

so that we can establish that $\varphi(n) \geq \frac{\sqrt{n}}{2}$.

Hence

$$n > 1 \Rightarrow \forall n \sum_{\substack{d | P_{\sqrt{n}} \\ d | 2 \cdot n}} \mu(d) \cdot \frac{n}{d} \geq \varphi(2 \cdot n) \geq \frac{\sqrt{2 \cdot n}}{2}$$

and

$$n > 1 \Rightarrow \forall n \sum_{d | P_{\sqrt{n}}} \frac{\mu(d)}{d} = \prod_{p \leq \sqrt{n}} \left(1 - \frac{1}{p}\right) = \prod_{\substack{p \leq \sqrt{n} \\ p | 2 \cdot n}} \left(1 - \frac{1}{p}\right) \cdot \prod_{\substack{p \leq \sqrt{n} \\ p \nmid 2 \cdot n}} \left(1 - \frac{1}{p}\right) \geq$$

$$\frac{\varphi(2 \cdot n)}{n} \cdot \frac{1}{\sqrt{n}} \geq \frac{\sqrt{2 \cdot n}}{2 \cdot n} \cdot \frac{1}{\sqrt{n}} = \frac{1}{\sqrt{2 \cdot n}}$$

Theorem n. 4

Theorem n. 4 Let $n > 1$ be an integer. Then

$$\forall n \exists p \in P \cup \{1\} \exists q \in P : p + q = 2n$$

Preamble.

The statement of Theorem n.4 can be rewritten in the form

$$n > 1 \rightarrow \forall n \; G(2n) \geq 1$$

so that a way to prove Theorem n.4 is that to prove that G(2n) is an superiorly unbound function.

Although the condition that G(2n) is an superiorly unbound function is not necessary.

In fact to obtain this result it is enough to prove that there exist two discrete functions g(2n), h(2n) and a real constant $0 < \varepsilon < 1$ such that

$$n > 1 \rightarrow G(2n) \geq g(2n) \quad (1)$$

$$n > 1 \rightarrow \forall n \; h(2n) > 1 \quad (2)$$

$$n > 1 \rightarrow \forall n \; |g(2n) - h(2n)| \leq (h(2n))^{\varepsilon} \quad (3)$$

Observe that, given (2), if $n > 1 \rightarrow \exists n \; g(2n) = 0$ then (3) is violated.

Preliminary observation to the proof of the Corollary to Theorem n.4

Set

$$A = \{(p,q) \in P \times P : p+q = 2n \wedge p < \sqrt{2n} < q < 2n\}$$

$$B = \{(p,q) \in P \times P : p+q = 2n \wedge \sqrt{2n} < p, q < 2n\}$$

$$g(2n) = |S| = \left|\{j \in \overline{n} : \forall p \in \overline{[\sqrt{2n}]} \cap P \; j \not\equiv n \bmod p \wedge j \not\equiv n \bmod p\}\right|$$

$$P_1^{\sqrt{n}} = \{p \in P : 1 < p \leq \sqrt{n}\}, \; P_{\sqrt{n}}^{\sqrt{2 \cdot n}} = \{p \in P : \sqrt{n} < p \leq \sqrt{2 \cdot n}\}$$

Observe now that the primes $q \in P_{\sqrt{n}}^{\sqrt{2 \cdot n}}$ do not concur in the determination of B and the primes $r \in P_1^{\sqrt{n}}$ concur twice in the determination of B iff

$$B = \{(p,p) \in P \times P : p + p = 2 \cdot n\}.$$

Now we can focus our attention on two functions. Set

$$h_1(n) = \frac{e^{-\gamma}}{\ln \sqrt{n}} \cdot n \cdot \prod_{\substack{p \leq \sqrt{2 \cdot n} \\ p \nmid 2 \cdot n}} \left(1 - \frac{1}{p}\right), \; h_2(n) = \frac{e^{-\gamma}}{\ln \sqrt{2 \cdot n}} \cdot n \cdot \prod_{\substack{p \leq \sqrt{2 \cdot n} \\ p \nmid 2 \cdot n}} \left(1 - \frac{1}{p}\right)$$

to observe that

$$n \geq 8 \Rightarrow \forall n \; h_1(n), h_2(n) > 1$$

Observe also that

$$\prod_{\substack{p \leq \sqrt{2 \cdot n} \\ p \nmid 2 \cdot n}} \left(1 - \frac{1}{p}\right) > \prod_{p \leq \sqrt{2 \cdot n}} \left(1 - \frac{1}{p}\right)$$

so that Theorem n.3 apply. Hence

$$h_1(n) + 1 = \frac{e^{-\gamma}}{\ln \sqrt{n}} \cdot n \cdot \prod_{\substack{p \leq \sqrt{2 \cdot n} \\ p \nmid 2 \cdot n}} \left(1 - \frac{1}{p}\right) + 1 > \frac{e^{-2\gamma} \cdot n}{\ln \sqrt{n} \cdot \ln \sqrt{2 \cdot n}}$$

and

$$h_2(n) + 1 = \frac{e^{-\gamma}}{\ln \sqrt{2 \cdot n}} \cdot n \cdot \prod_{\substack{p \leq \sqrt{2 \cdot n} \\ p \nmid 2 \cdot n}} \left(1 - \frac{1}{p}\right) + 1 > \frac{e^{-2\gamma} \cdot n}{\ln \sqrt{2 \cdot n} \cdot \ln \sqrt{2 \cdot n}}$$

so that for $n > 1$ $h_1(n), h_2(n)$ are increasing functions.

Set

$$\theta_0 = \prod_{p|2n, p\leq\sqrt{2n}} p, \quad \theta_1 = \prod_{p\backslash 2n, p\leq\sqrt{n}} p, \quad \theta_2 = \prod_{p\backslash 2n, p\leq\sqrt{2n}} p$$

$$\pi_0 = \prod_{p|\theta_0}(1-\frac{1}{p}), \quad \pi_1 = \prod_{p|\theta_1}(1-\frac{1}{p}), \quad \pi_2 = \prod_{p|\theta_2}(1-\frac{1}{p}),$$

$$\lambda_n \in]\pi_1^{-2}, \pi_2^{-2}[$$

and suppose that in the general cases

$$n>1 \wedge P_{\sqrt{n}}^{\sqrt{2\cdot n}} \neq \emptyset \Rightarrow \forall n \exists \lambda_n \; g(2\cdot n) = n\cdot\pi_0\cdot\pi_1^2\cdot\lambda_n\cdot\pi_2^2$$

$$n>1 \wedge P_{\sqrt{n}}^{\sqrt{2\cdot n}} = \emptyset \wedge 1<\varepsilon<2 \Rightarrow \forall n \exists \varepsilon \; g(2\cdot n) = n\cdot\pi_0\cdot\pi_1^\varepsilon = n\cdot\pi_0\cdot\pi_2^\varepsilon$$

By Theorem n.3 we have

$$|n\cdot\pi_0\cdot\pi_1^2\cdot\lambda_n\cdot\pi_2^2 - \frac{e^{-\gamma}}{\ln\sqrt{2\cdot n}}\cdot n\cdot\pi_1^2\cdot\lambda_n\cdot\pi_2| < \frac{n\cdot\pi_1^2\cdot\lambda_n\cdot\pi_2}{\left(\ln\sqrt{2\cdot n}\right)^2}$$

$$|n\cdot\pi_0\cdot\pi_2^\varepsilon - \frac{e^{-\gamma}}{\ln\sqrt{n}}\cdot n\cdot\pi_2^{\varepsilon-1}| < \frac{n\cdot\pi_2^{\varepsilon-1}}{\left(\ln\sqrt{n}\right)^2}$$

and

$$|g(2\cdot n) - \frac{e^{-\gamma}}{\ln\sqrt{2\cdot n}}\cdot n\cdot\pi_1^2\cdot\lambda_n\cdot\pi_2| < \frac{n\cdot\pi_1^2\cdot\lambda_n\cdot\pi_2}{\left(\ln\sqrt{2\cdot n}\right)^2}$$

$$|g(2\cdot n) - \frac{e^{-\gamma}}{\ln\sqrt{n}}\cdot n\cdot\pi_2^{\varepsilon-1}| < \frac{n\cdot\pi_2^{\varepsilon-1}}{\left(\ln\sqrt{n}\right)^2}$$

so that to prove that

$$n>1 \Rightarrow \forall n \; g(2\cdot n) \geq 1$$

we can exploit a proof by contradiction.

(Supposing that exist an $n>1$ such that $g(2\cdot n)=0$ we get a contradiction).

It is now plain what our aim in the general case.

I.e. setting n=10 we have G(2·n)=|{(1,19),(3,17),(7,13)}|,

$\pi_0 = \frac{1}{2}$, $\pi_1^2 \cdot \pi_2^2 = \frac{16}{81}$ so that $\lambda = \frac{9}{4}$ and

$G(2\cdot n) = 3 > \frac{20}{9} = n\cdot\lambda\cdot\pi_0\cdot\pi_1^2\cdot\pi_2^2$.

Proof of Theorem n. 4. Set

$$A = \{(p,q) \in P \times P : p+q = 2n \wedge p < \sqrt{2n} < q < 2n\}$$

$$B = \{(p,q) \in P \times P : p+q = 2n \wedge \sqrt{2n} < p,q < 2n\}$$

$$g(2n) = |S| = |\{j \in \overline{n} : \forall p \in \overline{[\sqrt{2n}]} \cap P \; j \neq n \bmod p \wedge j \neq n \bmod p\}|$$

$$\theta_0 = \prod_{p|2n,\, p \leq \sqrt{2n}} p, \quad \theta_1 = \prod_{p \backslash 2n,\, p \leq \sqrt{n}} p, \quad \theta_2 = \prod_{p \backslash 2n,\, p \leq \sqrt{2n}} p$$

$$\pi_0 = \prod_{p|\theta_0}\left(1-\frac{1}{p}\right), \quad \pi_1 = \prod_{p|\theta_1}\left(1-\frac{1}{p}\right), \quad \pi_2 = \prod_{p|\theta_2}\left(1-\frac{1}{p}\right),$$

The fact that

$$n > 4 \wedge 0 < \varepsilon < 1 \Rightarrow \forall n \exists \varepsilon \; \sum_{d|\theta_2} \mu(d) \cdot \left[\frac{n}{d}\right] + \varepsilon$$

$$= \sum_{d|\theta_2} \mu(d) \cdot \frac{n}{d} = n \cdot \prod_{p|\theta_2}\left(1-\frac{1}{p}\right),$$

(Corollaries to Theorem n.2)

will grant us to exploit Eratosthenes-Legendre Theorem in the following of the proof.

Firstly suppose that $P_{\sqrt{n}}^{\sqrt{2 \cdot n}} \neq \emptyset$.

Since $|B|$ is $g(2n)-1$ if $2n-1 \in P$ and $g(2n)$ otherwise
and since
$G(2n)=g(2n)+|A| \geq |B|+|A|$
to show that
$G(2n) \geq 1$
we can prove that $g(2n)$ is a superiorly unbound function in n.

Firstly we can prove that $n \cdot \pi_0 \cdot \pi_2^2$ is a minorant for $g(2n)$.

Define the set
$E = \{(a,b) \in \{1,2,...n-1,n\} \times \{n, n+1,...2n-1, 2n\} : a+b = 2n \wedge (a, \theta_0) = 1 = (b, \theta_0)\}$
in order to observe - since $(k,n)=1 \Rightarrow (n-k,n)=1$ - that
$n \cdot \pi_0 = |E|$

Now, since in the definition of S for every n-j
there exists a n+j such that $(n-j)+(n+j)=2n$
and such that $(n-j, n+j) \in P \times P$, $(n-j, \theta_1) = 1 = (n+j, \theta_2)$
it is clear that we can prove
$n \cdot \pi_0 \cdot \pi_2^2 < |S| \leq n \cdot \pi_0 \cdot \pi_1^2$.

Namely we can prove that
$n > 1 \Rightarrow (\forall n \exists \lambda \ \pi_2^2 < \lambda^{-1} < \pi_1^2 \Rightarrow |S| = n \cdot \pi_0 \cdot \lambda \cdot \pi_2^2 \cdot \pi_1^2)$

In fact suppose that
$|S| = n \cdot \pi_0 \cdot \lambda \cdot \pi_2^2 \cdot \pi_1^2 \wedge \lambda^{-1} \leq \pi_2^2 \wedge B \neq \{(p,p) \in P \times P : p+p = 2 \cdot n\}$ (1)
Condition (1) shows that the primes in the set
$\{p \in P : \sqrt{n} < p \leq \sqrt{2 \cdot n} \wedge p \setminus 2 \cdot n\}$
do not concur in the determination of S,
contradicting Eratosthenes-Legendre Theorem.
The contradiction we have reached shows that
$n > 1 \Rightarrow (\forall n \exists \lambda \ \pi_2^2 < \lambda^{-1} \Rightarrow |S| = n \cdot \pi_0 \cdot \lambda \cdot \pi_2^2 \cdot \pi_1^2)$

Suppose also that

$$|S| = n \cdot \pi_0 \cdot \lambda \cdot \pi_2^2 \cdot \pi_1^2 \wedge \lambda^{-1} \geq \pi_1^2 \wedge B \neq \{(p,p) \in P \times P : p + p = 2 \cdot n\} \quad (2)$$

Condition (2) shows that the primes in the set

$$\{p \in P : 1 < p \leq \sqrt{n} \wedge p \mid 2 \cdot n\}$$

do not concur proportionately in the determination of S

(to understand the meaning of

"do not concur proportionately" see the previous part of the proof).

The contradictions we have reached show that

$$n > 1 \Rightarrow (\forall n \exists \lambda \ \pi_2^2 < \lambda^{-1} < \pi_1^2 \Rightarrow |S| = n \cdot \pi_0 \cdot \lambda \cdot \pi_2^2 \cdot \pi_1^2)$$

Hence $\pi_0 \cdot \pi_2^2$, $\pi_0 \cdot \pi_1^2$

are respectively a minorant and an upper bound for $\dfrac{g(2 \cdot n)}{n} = \pi_0 \cdot \lambda \cdot \pi_2^2 \cdot \pi_1^2$

$$\left(\begin{array}{l} \pi_0 \cdot \pi_1^2 \text{ is an upper bound for } \pi_0 \cdot \lambda \cdot \pi_2^2 \cdot \pi_1^2 \\ \text{since } g(4) = 1 \Rightarrow \dfrac{g(4)}{2} = \dfrac{1}{2} \cdot 1 = \prod_{\substack{p \mid 2 \\ p \leq 2}} \left(1 - \dfrac{1}{p}\right) \cdot \prod_{\substack{p \nmid 2 \\ p \leq 2}} \left(1 - \dfrac{1}{p}\right) \end{array} \right)$$

Therefore for every n greater than 1 there exists a real variable λ such that

$\pi_2^2 < \lambda \cdot \pi_2^2 \cdot \pi_1^2 < \pi_1^2$

and such that

$g(2n) = n \cdot \pi_0 \cdot \lambda \cdot \pi_2^2 \cdot \pi_1^2$

Now suppose that $P_{\sqrt{n}}^{\sqrt{2 \cdot n}} = \emptyset$. We have $\pi_2 = \pi_1$.

Let ε be a real number such that $|\varepsilon| \in {]}0,1[$.

Firstly I shall prove that if $\dfrac{g(2 \cdot n)}{n \cdot \pi_0} = \pi_2^{2+\varepsilon}$ then ε has negative sign,

then I shall prove that for any ε we have $1 \leq n \cdot \pi_0 \cdot \pi_2^{2-|\varepsilon|}$.

To start observe that n for which the set

$\left\{ p \in P : 1 < p \leq \dfrac{\sqrt{n}}{2} \wedge p \nmid 2 \cdot n \right\}$ is the empty set or the set $\left\{ p \in P : \dfrac{\sqrt{n}}{2} < p \leq \sqrt{n} \wedge p \nmid 2 \cdot n \right\}$

is the empty set are all multiple of 6 and are all included between 49 and 100 or are less than 49.

In other words we shall prove that in the case $P_{\sqrt{n}}^{\sqrt{2 \cdot n}} = \emptyset$ we have that

$n \in [50,100] \wedge 6 \nmid n \wedge \delta \in {]}0,1[\wedge P_{\sqrt{n}}^{\sqrt{2 \cdot n}} = \emptyset \wedge B \neq \{(p,p) \in PXP : p + p = 2 \cdot n\}$

$\Rightarrow \forall n \exists \delta \; n \cdot \pi_0 \cdot \pi_2^{1+\delta} = g(2 \cdot n)$.

To realize the proof we have to split the set $\left\{ p \in P : 1 < p \leq \sqrt{n} \wedge p \nmid 2 \cdot n \right\}$ in two subsets, say A,B.

Then we have to show that the subset "more heavy" concurs more than the subset "more light" in determination of $g(2 \cdot n)$.

Define then

$$\delta^+ = \dfrac{\left|\left\{ p \in P : \dfrac{\sqrt{n}}{2} < p \leq \sqrt{n} \wedge p \nmid 2 \cdot n \right\}\right|}{\left|\left\{ p \in P : 1 < p \leq \sqrt{n} \wedge p \nmid 2 \cdot n \right\}\right|}, \; \delta^- = \dfrac{\left|\left\{ p \in P : 1 < p \leq \dfrac{\sqrt{n}}{2} \wedge p \nmid 2 \cdot n \right\}\right|}{\left|\left\{ p \in P : 1 < p \leq \sqrt{n} \wedge p \nmid 2 \cdot n \right\}\right|}$$

$$S^+(n) = \left\{ \begin{array}{l} (m,r) \in \left\{ \left[\dfrac{n}{4}\right]+1, \left[\dfrac{n}{4}\right]+2, \ldots n-\left[\dfrac{n}{4}\right]-2, n-\left[\dfrac{n}{4}\right]-1 \right\} \cap PX \\ \left\{ \left[\dfrac{n}{4}\right]+1, \left[\dfrac{n}{4}\right]+2, \ldots n-\left[\dfrac{n}{4}\right]-2, n-\left[\dfrac{n}{4}\right]-1 \right\} \cap P : \; m+r = 2 \cdot n \end{array} \right\},$$

$$S^-(n) = \left\{ (m,r) : m \in \left\{1,2,\ldots \left[\dfrac{n}{4}\right]\right\} \cap P \wedge r \in \left\{ n-\left[\dfrac{n}{4}\right], n-\left[\dfrac{n}{4}\right]+1, \ldots n \right\} \cap P : m+r = 2 \cdot n \right\},$$

$$S_0(n) = \left\{ m \in \left\{1,2,\ldots \left[\dfrac{n}{4}\right]\right\} \cap P : \exists r \in \left\{ n-\left[\dfrac{n}{4}\right], n-\left[\dfrac{n}{4}\right]+1, \ldots n \right\} \cap P : m+r = 2 \cdot n \right\}$$

Observe now that for $n \geq 50$

$$\prod_{\substack{1 < p \leq \frac{\sqrt{n}}{2} \\ p \nmid 2 \cdot n}} \left(1 - \dfrac{1}{p}\right) < \prod_{\substack{\frac{\sqrt{n}}{2} < q \leq \sqrt{n} \\ q \nmid 2 \cdot n}} \left(1 - \dfrac{1}{q}\right)$$

so that the set $\left\{ p \in P : 1 < p \leq \dfrac{\sqrt{n}}{2} \wedge p \nmid 2 \cdot n \right\}$ is "more heavy" than the set

$\left\{ p \in P : \dfrac{\sqrt{n}}{2} < p \leq \sqrt{n} \wedge p \nmid 2 \cdot n \right\}$.

The set $\left\{p \in P : \dfrac{\sqrt{n}}{2} < p \leq \sqrt{n} \wedge p \nmid 2 \cdot n\right\}$ does not concur in the determination of $S_0(n)$,

$\left(\text{In fact } \left[\dfrac{n}{4}\right]^{\frac{1}{2}} \text{ is less than any element of the set:} \left\{p \in P : \dfrac{\sqrt{n}}{2} < p \leq \sqrt{n} \wedge p \nmid 2 \cdot n\right\}\right)$

so that the set $\left\{p \in P : \dfrac{\sqrt{n}}{2} < p \leq \sqrt{n} \wedge p \nmid 2 \cdot n\right\}$ concurs less than the set

$\left\{p \in P : 1 < p \leq \dfrac{\sqrt{n}}{2} \wedge p \nmid 2 \cdot n\right\}$ in the determination of $S^-(n)$,

but the same one concurs fully in the determination of $S^+(n)$. Therefore the comprehensive contribution of the set $\left\{p \in P : 1 < p \leq \dfrac{\sqrt{n}}{2} \wedge p \nmid 2 \cdot n\right\}$ to the determination of the quantity

$|S^-(n)| + |S^+(n)|$ is greater than the contribution of the set $\left\{p \in P : \dfrac{\sqrt{n}}{2} < p \leq \sqrt{n} \wedge p \nmid 2 \cdot n\right\}$

to the determination of the quantity $|S^-(n)| + |S^+(n)|$.

The fractions δ^+, δ^- do not give the relative contributions of the sets
I have cited above to the determination of the quantity $|S^-(n)| + |S^+(n)|$.

To esteeme the sum of these relative contributions, say δ, we can set
$\delta^+ + \delta^- < \delta \leq \delta^+ + 2 \cdot \delta^-$
Therefore δ is greater than 1 and less than 2.
Hence
$n \in [50, 100] \wedge 6 \nmid n \wedge \delta \in]1, 2[\wedge P_{\sqrt{n}}^{\sqrt{2 \cdot n}} = \emptyset \wedge B \neq \{(p, p) \in P \times P : p + p = 2 \cdot n\}$
$\Rightarrow \forall n \exists \delta \; n \cdot \pi_0 \cdot \pi_2^\delta = g(2 \cdot n)$.

Now the completion of the proof of Theorem n.4.

Notice that by Theorem n.3

$$\left| \prod_{p \leq \sqrt{2n}} (1 - \frac{1}{p}) - \frac{e^{-\gamma}}{\ln \sqrt{2 \cdot n}} \right| \leq \frac{1}{(\ln \sqrt{2 \cdot n})^2}$$

and

$$\left| \prod_{p \leq \sqrt{n}} (1 - \frac{1}{p}) - \frac{e^{-\gamma}}{\ln \sqrt{n}} \right| \leq \frac{1}{(\ln \sqrt{n})^2}$$

Multiplying LHS and RHS by $n \cdot \lambda \cdot \pi_2 \cdot \pi_1^2$ in first case and by $n \cdot \pi_2^{\delta-1}$ ($\delta \in]1,2[$) in the second case, we get

$$\left| g(2n) - \frac{e^{-\gamma} n \cdot \lambda \cdot \pi_2 \cdot \pi_1^2}{\ln \sqrt{2 \cdot n}} \right| \leq \frac{n \cdot \lambda \cdot \pi_2 \cdot \pi_1^2}{(\ln \sqrt{2 \cdot n})^2}$$

and

$$\left| g(2n) - \frac{e^{-\gamma} \cdot n \cdot \pi_2^{\delta-1}}{\ln \sqrt{n}} \right| \leq \frac{n \cdot \pi_2^{\delta-1}}{(\ln \sqrt{n})^2}$$

The two inequalities

$$\left| g(2n) - \frac{e^{-\gamma} n \cdot \lambda \cdot \pi_2 \cdot \pi_1^2}{\ln \sqrt{2 \cdot n}} \right| \leq \frac{n \cdot \lambda \cdot \pi_2 \cdot \pi_1^2}{(\ln \sqrt{2 \cdot n})^2},$$

$$\left| g(2n) - \frac{e^{-\gamma} \cdot n \cdot \pi_2^{\delta-1}}{\ln \sqrt{n}} \right| \leq \frac{n \cdot \pi_2^{\delta-1}}{(\ln \sqrt{n})^2}$$

by what we have proven above,

allow us to give rise to a proof by contradiction.

Recall now that in the Preliminary Observation to the proof of Theorem n.4 we have set

$$h_1(n) = \frac{e^{-\gamma}}{\ln \sqrt{n}} \cdot n \cdot \prod_{\substack{p \leq \sqrt{2n} \\ p \nmid 2 \cdot n}} \left(1 - \frac{1}{p}\right), h_2(n) = \frac{e^{-\gamma}}{\ln \sqrt{2 \cdot n}} \cdot n \cdot \prod_{\substack{p \leq \sqrt{2n} \\ p \nmid 2 \cdot n}} \left(1 - \frac{1}{p}\right)$$

and we have found that for $n \geq 8$ $h_1(n), h_2(n) > 1$. Further we have found that $h_1(n), h_2(n)$ are increasing functions in $n \in [2, \infty[$.

There are the conditions for the completion of the proof by the Preamble.

Observe that if $g(2 \cdot n)$ was 0 for $n > 30$ then the inequality

$$\left| g(2n) - \frac{e^{-\gamma} n \cdot \lambda \cdot \pi_2 \cdot \pi_1^2}{\ln \sqrt{2 \cdot n}} \right| \leq \frac{n \cdot \lambda \cdot \pi_2 \cdot \pi_1^2}{(\ln \sqrt{2 \cdot n})^2}$$

or the inequality

$$\left| g(2n) - \frac{e^{-\gamma} n \cdot \pi_2^{\delta-1}}{\ln \sqrt{n}} \right| \leq \frac{n \cdot \pi_2^{\delta-1}}{(\ln \sqrt{n})^2}$$

would be violated.

This observation concludes the proof.

Final Observations to the proof of Theorem n.4

Observation n.1.

The number of positive integers for whom $P_{\sqrt{n}}^{\sqrt{2n}} = \emptyset$ is finite.

Observation n.2.

Bertrand's Postulate does not apply to the set

$$\left\{ p \in P : \frac{\sqrt{n}}{2} < p \leq \sqrt{n} \wedge p \nmid 2 \cdot n \right\}$$

Conclusion

Let A be the set of prime numbers r such that

- r divides 2n;
- $r \leq \sqrt{n}$.

Let B be the set of prime numbers p such that

- p does not divide 2n;
- $p \leq \sqrt{n}$.

Let C be the set of prime numbers q such that

- q does not divide 2n;
- $\sqrt{n} < q \leq \sqrt{2n}$.

Let R(2n) be a function that takes values in the interval

[0,∞] and such that for any n R(2n) ≤ g(2n)

n, |A|, |B|, |C| and R(2n) concur to determinate the number of elements of goldbachian pairs,

by Eratosthenes-Legendre Theorem so that n, |A|, |B|, |C| and R(2n) concur to determinate the number of goldbachian pairs

Thus by Eratosthenes-Legendre Theorem we have deduced the functional form that

g(2n) has. g(2n) is a function in n, |A|, |B| and |C|, so that G(2n)

is a function in n, |A|, |B|, |C| and R(2n). In formulae

G(2n)=g(2n)+R(2n)=G(n, |A|, |B|, |C| and R(2n))

To reject my proof is equal to reject Eratosthenes-Legendre Theorem, so that to reject my proof is equal to reject the Foundations of Mathematics.

Appendix

Lemma.

$$n \geq 1 \Rightarrow \forall n \ \sum_{d|n} \mu(d) = \left[\frac{1}{n}\right]$$

Proof of the Lemma.

Since we already know that $\mu(1) = 1$, suppose that $n > 1$.

We have to prove that

$$n > 1 \Rightarrow \forall n \ \sum_{d|n} \mu(d) = 0$$

By Theorem n.2 we have

$$n > 1 \Rightarrow \forall n \ \sum_{d|n} \mu(d) = \prod_{p|n}(\mu(1) + \mu(p)) = \prod_{p|n}(1-1) = 0$$

as desired.

Teorem n.5.

$$n \geq 1 \Rightarrow \forall n \ \varphi(n) = \sum_{d|n} \mu(d) \cdot \frac{n}{d}$$

Proof of Theorem n.5.

Since

$$n \geq 1 \Rightarrow \forall n \ \varphi(n) = \sum_{k=1}^{n} \left[\frac{1}{(n,k)}\right]$$

by the Lemma we have

$$n \geq 1 \Rightarrow \forall n \ \varphi(n) = \sum_{k=1}^{n} \left[\frac{1}{(n,k)}\right] = \sum_{k=1}^{n} \sum_{d|(n,k)} \mu(d) = \sum_{k=1}^{n} \sum_{\substack{d|n \\ d|k}} \mu(d)$$

Observe now that

$$k = qd \Rightarrow \forall k \forall d \exists q \left(1 \leq k \leq n \Leftrightarrow 1 \leq q \leq \frac{n}{d}\right)$$

Hence

$$n \geq 1 \Rightarrow \forall n \ \varphi(n) = \sum_{k=1}^{n} \sum_{\substack{d|n \\ d|k}} \mu(d) = \sum_{d|n} \sum_{q=1}^{\frac{n}{d}} \mu(d) = \sum_{d|n} \mu(d) \sum_{q=1}^{\frac{n}{d}} 1 = \sum_{d|n} \mu(d) \cdot \frac{n}{d}$$

as desired.

Theorem n.6.

$$n \geq 1 \Rightarrow \forall n \ \varphi(n) = n \cdot \prod_{p|n} \left(1 - \frac{1}{p}\right)$$

Proof of Theorem n.6.

Set

$$\prod_{p \in \varnothing} \left(1 - \frac{1}{p}\right) = 1$$

(the empty product is 1) to obtain $\varphi(1) = 1$.

Now suppose $n > 1$.

By the first corollary to Theorem n.2 we have

$$n \geq 1 \Rightarrow \forall n \ n \cdot \prod_{p|n}\left(1 - \frac{1}{p}\right) = n \cdot \sum_{d|n} \frac{\mu(d)}{d} = \sum_{d|n} \mu(d) \cdot \frac{n}{d}$$

By Theorem n.5 we have

$$n \geq 1 \Rightarrow \forall n \ \varphi(n) = \sum_{d|n} \mu(d) \cdot \frac{n}{d}$$

so that

$$n \geq 1 \Rightarrow \forall n \ \varphi(n) = n \cdot \prod_{p|n}\left(1 - \frac{1}{p}\right)$$

as desired.

Bertrand's Postulate. Let $n > 1$ be an integer.
Then between n and 2n there exists at least a prime number.

Abel summation formula. Suppose $\{a_n\}_{n=1}^{\infty}$ is a sequence of complex number and f is a continuously differentiable function in $[1, x]$. Set
$$A(t) = \sum_{n \leq t} a_n$$
Then
$$\sum_{n \leq x} a_n \cdot f(n) = A(x) \cdot f(x) - \int_1^x A(t) \cdot f'(t) \, dt$$

Proof. Suppose x is a natural number.
We write the LHS as:
$$\sum_{n \leq x} a_n \cdot f(n) = \sum_{n \leq x} \{A(n) - A(n-1)\} \cdot f(n) =$$
$$\sum_{n \leq x} A(n) \cdot f(n) - \sum_{n \leq x-1} A(n) \cdot f(n+1) =$$
$$A(x) \cdot f(x) - \sum_{n \leq x-1} A(n) \cdot \int_n^{n+1} f'(t) \, dt =$$
$$A(x) \cdot f(x) - \sum_{n \leq x-1} \int_n^{n+1} A(t) \cdot f'(t) \, dt =$$
$$A(x) \cdot f(x) - \int_1^x A(t) \cdot f'(t) \, dt$$

since $A(t)$ is a step function.
We have proven the result if x is an integer.
If x is not an integer observe that we have
$$\{f(x) - f([x])\} \cdot A(x) = \int_{[x]}^x A(t) \cdot f'(t) \, dt$$
which completes the proof.

Define now

$$\Lambda(n) = \begin{cases} \log p \text{ if } n = p^\alpha, \ \alpha \geq 1 \\ 0 \text{ otherwise} \end{cases}$$

and

$$\psi(n) = \sum_{n \leq x} \Lambda(n)$$

so that

$$\sum_{e \leq x} \psi\left(\frac{x}{e}\right) = \sum_{d \cdot e \leq x} \Lambda(d) = \sum_{n \leq x} \log n$$

Since if $a_0 \geq a_1 \geq a_2 \geq \ldots$ is a decreasing sequence of real numbers tending to 0 then

$$a_0 - a_1 \leq \sum_{n=0}^{\infty} (-1)^n a_n \leq a_0 - a_1 + a_2$$

so that

$$\psi(x) - \psi\left(\frac{x}{2}\right) \leq \sum_{n=0}^{\infty} (-1)^n \psi\left(\frac{x}{n}\right) \leq \psi(x) - \psi\left(\frac{x}{2}\right) + \psi\left(\frac{x}{3}\right)$$

Observe that

$$\psi(x) - 2 \cdot \psi\left(\frac{x}{2}\right) > 0 \Rightarrow \forall x \exists p \ \frac{x}{2} < p < x$$

By Abel Summation Formula, setting $A(t) = [t], f(t) = \log t$ and $\delta(t) = \frac{[t]}{[t]+\{t\}} < 1$ we have

$$\sum_{n \leq x} \log n = [x] \cdot \log x - \int_1^x \frac{[t]}{t} dt = [x] \cdot \log x - \delta(x) \cdot x + 1 = x \cdot \log x - x + O(\log x)$$

Hence, we have

$$\psi(x) - 2 \cdot \psi\left(\frac{x}{2}\right) = x \cdot \log x - x - x \cdot \log \frac{x}{2} + x + O(\log x) = x \cdot \log 2 + O(\log x) \quad (1)$$

and

$$A \leq \frac{|\psi(x) - 2 \cdot \psi\left(\frac{x}{2}\right) - x \cdot \log 2|}{\log x} \leq B \quad (2)$$

so that if we prove that $A > -\frac{\log 2^x}{\log x}$ then we prove that $\psi(x) - 2 \cdot \psi\left(\frac{x}{2}\right) > 0.$

A way to conclude the proof, as we are going to see, is that to prove that $A \geq -\frac{\log 2^x}{\log 2 \cdot x}$.

By Stirling's Theorem we have
$$x \log x - x \leq \log x! = \psi(x)$$
and
$$2 \cdot \psi\left(\frac{x}{2}\right) = 2 \cdot \log\left(\left(\frac{x}{2}\right)!\right) \leq x \log x - x - x \log 2 + \log x + \log 2$$
so that
$$\psi(x) - 2 \cdot \psi\left(\frac{x}{2}\right) \geq x \log 2 - \log x - \log 2 > 0, \ x \geq 4$$

Bibliography

David M. Burton, Elementary Number Theory

Gareth A. Jones and J. Mary Jones, Elementary Number Theory

Murray Spiegel, Advanced Calculus

Tom M. Apostol, Introduction to Analytic Number Theory

Ivan Niven, Herbert S. Zuckerman and Hugh L. Montgomery, An Introduction to the Theory of Numbers

M. Ram Murty, Problems in Analytic Number Theory

Gerard Tenenbaum, Introduction to Analytic and Probabilistic Number Theory

Yuan Wang, The Goldbach Conjecture